ZANY BRAINY ANIMALS

HOW ANIMALS DEFEND THEMSELVES

WAYLAND

First published in Great Britain in 2024 by Wayland
Copyright © Hodder and Stoughton, 2024

Editor: Elise Short
Designer: Rachel Lawston

HB ISBN: 978 1 5263 2339 2
PB ISBN: 978 1 5263 2340 8

Printed and bound in China

Wayland, an imprint of
Hachette Children's Group
Part of Hodder and Stoughton
Carmelite House
50 Victoria Embankment
London EC4Y 0DZ
An Hachette UK Company
www.hachette.co.uk
www.hachettechildrens.co.uk

MIX
Paper from
responsible sources
FSC
www.fsc.org
FSC® C104740

Picture credits:
Alamy: Ethan Daniels13t,Laura Romin & Larry Dalton 15c.
Shutterstock: Aastels 10t, Allexxandar13b b/g, Evgeny Ayupov
9tr, Brent Barnes 20tr, Bear Fotos 20cl, Audrey Snider-Bell
26t,Randy Bjorklund 8br, Mark Bridger 10b, Yulia Burkarova
8t, Chainfoto24 4c,18c, Craig Dingle 22t, Ashton Ea 6,
Foreverhappyfr cover b,Tyler Fox 8bl, GypsyPictureShow 23br,
Happy monkey 15b, Eric Isselee 19b, Rosa Jay 5b, IrinaK 24,
Irin-k 11tl, Randi Kayu 19tr, Prahab Louilarpprasert 9bl, Fabio
Maffei 16t, Benny Marty 26bl, Liz Miller 28t, Jay Ondreika 25t,
Pakhnyushchy 17c,Vadim Petrakov 21t, 21b, Pinosub 7b,PRILL
28b, SoFlo Shots 5t, 23tr, Nickolay Stanev 29t,David Pineda
Svenske 23tl, Johan Swanepoel 14bl,Torook 27b, Wildestanimal
29b. Wikimedia Commons: Dan Schofield 12b, CCA 4.0
International.

CONTENTS

HOW DO ANIMALS DEFEND THEMSELVES?

For lots of animals, life can sometimes be dangerous. So, to keep themselves safe, animals have to defend themselves. There are many different ways of doing this of course, from being **SUPER FAST** to **SUPER STRONG** and from being **SUPER SMART** to **SUPER SNEAKY!**

You'll never guess how I defend myself!

A tortoise's defences are rock solid! How do I get in?

In this book, we'll look at the fascinating ways in which animals defend themselves, getting one over on their enemies to make sure that they win the game of life.

TALES FROM THE LAB

Understanding animal behaviour is not an easy task. First, scientists watch animals living their life. That's how they get their ideas. Then they test these ideas by running experiments. If they're lucky, they'll find out something that no one has ever known about animals, which is the most exciting thing possible! In this book, you'll find many amazing discoveries about how animals defend themselves.

Don't get too close, I'm spiky!

CAMOUFLAGE

One of the best ways to avoid being discovered by a hungry predator is to blend into the background, a sneaky strategy known as camouflage.

Some animals, like the leafy sea dragon, take this to such extreme lengths that you might not see it even when you're looking straight at it. Part of the reason is that the sea dragon is exactly the same colour as the seaweed that it lives amongst. It can even change colour to make sure it blends in.

In addition to this, the sea dragon has evolved a **VERY STRANGE** shape. Sprouting from its body are all sorts of special growths that make it look less like a fish and more like seaweed.

Am I seaweed or am I a fish? Can you tell?

If I keep swaying, he'll never notice me.

Where's my lunch? That can't be it: it's just some floating seaweed!

Added to that, the sea dragon sways backwards and forwards in the water currents – again, just like seaweed.

All in all, the sea dragon is a **MASTER OF DISGUISE**, hiding in plain sight, safe in the knowledge that predators, such as sharks, can't eat it if they can't find it.

I'm hungry!

One of the weirdest kinds of camouflage is used by the lemon damsel. It's weird because these coral reef fish are bright yellow. But although they stand out to us, the kinds of fish that eat lemon damsels have very different vision to humans and, to their eyes, a lemon damsel seems a murky grey colour that blends into the background.

You think I'm yellow, but he thinks I'm grey! Ha!

7

MIMICRY AND MASQUERADE

There's more than one way to change your appearance. While camouflage works for some creatures, plenty of others pretend to be something they're not, by mimicking (copying) or masquerading (disguising themselves) as another animal.

Dave! Stop pooing everywhere!

Hey! That wasn't me!

The caterpillar of the swallowtail butterfly looks exactly like bird poo. This strategy works really well, because ...

WHO WANTS TO EAT POO?

Heehee! I'm not Dave's poo, I'm a caterpillar.

Some of the best masqueraders are the many different species of fly who pretend to be their own worst enemy: spiders. Snowberry flies, for instance, have patterns on their wings that look exactly like spider legs.

Spider or fly? Can you tell?

Uh-oh! Spider!

I see my dinner up ahead!

Oh yes, I can be a hairy spider, too!

Whenever the fly is approached by a spider, all it has to do is to stick its wings out at either side and – hey, presto – the fly now looks like a pretty big spider, complete with **EXTREMELY HAIRY LEGS.**

Completely fooled and suddenly in fear for its own life, the real spider turns and runs away, leaving the fly to enjoy its afternoon.

Not dinner! Too scary! I'm off!

TALES FROM THE LAB

For a long time, the patterns on some insects' wings were thought to simply be a way for creatures of the same species to recognise each other. That is, until clever zoologists (animal scientists) realised that they looked like something else: their predator! In experiments carried out on a type of fly, scientists wiped off the patterns on the flies' wings and found that predators didn't flee as often. Instead, they pounced!

DON'T EAT ME, I'M POISONOUS

If you're an animal small enough to be eaten, such as a little frog, one thing that might save you is if you taste awful or, better still, if you're poisonous.

The thing is, the frog doesn't want to actually be eaten for the predator to find out that it's the very opposite of delicious.

The frog wants to make sure it gets the message across **LOUD AND CLEAR** – before any chewing takes place.

Yummy frog!

No! Don't touch it, it's poisonous!

As a result, animals like the poison arrow frog go into advertising. They sport bright colours that make them stand out and tells any potential predator that it would be a **TERRIBLE IDEA** to eat them. And since they want to be sure that there's no mistake, different poisonous animals often feature the same colours.

A combination of yellow and black is a very popular kind of warning colour combo in the animal kingdom – think wasps, bees, snakes – the list goes on. These animals all benefit from a kind of universal signal that shouts out **'DON'T EAT ME – OR YOU'LL REGRET IT!'**

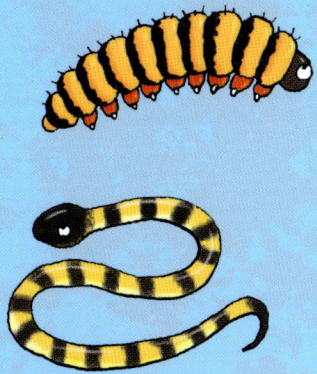

Strange though it might seem, some of the poisons that animals use are really valuable to us. Scientists have carefully developed medicines from these dangerous substances and some are now used to treat things like heart problems and diabetes, and even as a remedy for pain!

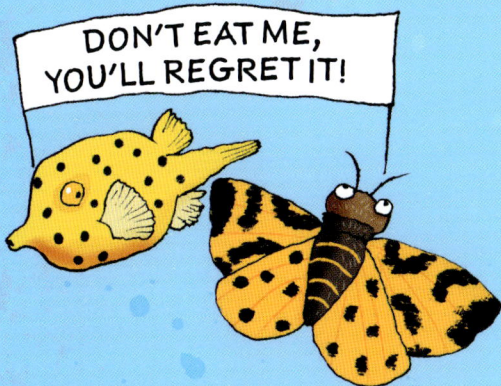

DON'T EAT ME, YOU'LL REGRET IT!

Yellow and black! Don't eat that!

HIDING

One of the simplest options available to animals that feel threatened is to hide. Lots of animals build shelters or dig burrows to keep themselves safe from harm. Thousands of years ago, our ancient ancestors lived in caves, partly out of fear of scary animals like **SABRE-TOOTHED TIGERS!**

I hope I'm safe in this cave!

But what if you're not very good at digging or making your own refuge? That's when animals have to get creative. One of the **WEIRDEST** examples of this is the pearlfish, which, when threatened, does something truly **BIZARRE**: it hides inside a sea cucumber's bottom.

It's dangerous around here! I must hide! A sea cucumber bum ahead! Perfect!

Sea cucumbers are often quite large, slow-moving animals that protect themselves using nasty-tasting or poisonous chemicals. Very few animals eat them, which is exactly why the pearlfish chooses them as its hiding place.

It's not known whether pearlfish think it's pretty ICKY inside a sea cucumber's bum, but one thing's for sure – it's better than getting eaten!

Although animals go into hiding when they're in danger, some animals carry their shelter around with them. Hermit crabs have quite soft bodies, so they use old snail shells to hide in, curling the back half of their body inside the spiralling shell.

Hey! Who said you could barge in?

Safe at last! It's cosy in here.

It's moving day! Which new home will I choose?

I'VE SEEN YOU!

Predators like to sneak up on their prey, stealthily creeping closer and closer, until they're almost on top of their victim before launching an attack. The **SECRET** to being a successful hunter is to make sure they're not spotted as they approach.

Most prey animals keep a wary eye out for exactly this kind of sly behaviour. If they detect approaching danger, they run off at top speed in the opposite direction.

Tonight's dinner should be wildebeest! Yum!

Running isn't always the best strategy though – it uses up precious energy and it means that you end up in a place you might not actually want to be. So, rather than run when they spot an advancing predator, animals like Thomson's gazelles simply let that predator know that they've seen it.

Lion in sight! Run away!

I've seen you, pesky lion!

In a behaviour known as 'stotting', a gazelle jumps into the air repeatedly, bouncing up and down like CRAZY until its predator gets the message. The crucial element of surprise has been lost and it won't be catching this gazelle today.

It's vitally important for prey animals to be able to spot when a predator is sneaking up on them. You might have noticed that animals such as rabbits or sheep have eyes on the sides of their head – this means they have the widest visual field and can see almost all the way around them.

I need to be able to see ahead and on both sides!

SURPRISE!

Some animals turn the tables on their attackers, giving them a nasty surprise that makes them think **TWICE**.

Ssss! Look at that easy lunch!

Is that a snake creeping up on me?

Imagine you're a snake, creeping up on what looks like a delicious and defenceless frog. *'This is too easy!'* you think to yourself as you slither closer and get ready to strike.

Eeek! What's that?!

Suddenly, the frog, a Cuyaba dwarf frog, does a **WEIRD BIT OF YOGA** that involves sticking its bum in the air. Best of all, the frog has what looks like a pair of great big eyes on its behind, so the snake now finds itself looking at what appears to be the face of a much larger animal.

All of this is too much for the snake, which changes its mind about attacking and scoots away, defeated by a **TOTALLY TERRIFYING FROG** with a face on its bum.

That face on my bum never lets me down!

So scary! That's put me off my lunch ...

I'm out of here. It sounds like there's a rattlesnake in there!

Hiss, hiss!

Rattle, rattle!

It's one thing to surprise a predator in broad daylight, but how can you do this if you live in the dark? The burrowing owl, which nests underground, has a neat solution. If a dangerous animal comes to the burrow, the owls make **hissing** and **rattling** noises that convince the intruder there's a deadly rattlesnake hiding within!

TOO BIG TO EAT

With a bit of sneaky trickery, a bite-sized animal can transform itself in an instant into something that's too hefty to handle.

Pufferfish are famous for this – it's where their name comes from. One moment this little fish might be swimming along, minding its own business, when suddenly a predator appears with a nasty glint in its eye.

Just the snack I was looking for!

Hmm. What a lovely day!

In the blink of an eye, the once small pufferfish drinks down a huge amount of water and expands up to three times its original size. Some puffer fish even have spikes!

The **MEGA-INFLATED** puffer is far too big for the predator to get its mouth around, so pufferfish is off the menu. It's a great defence, but don't try it yourself – it only works for pufferfish!

Yeesh! I can't eat that!

Gulp, gulp, gulp!

Hiss! I'm bigger than you think!

Have you ever seen a cat get really scared? it arches its back, its fur stands up and its tail goes bushy. Just like the pufferfish, this makes the cat seem bigger and more daunting to its enemies!

LET'S GET OUT OF HERE!

When danger threatens, the best option can be to just get out of the way. Lots of prey animals can run – or swim, or fly – out of trouble incredibly quickly. Other animals take this a step further and simply disappear.

Up, up and away!

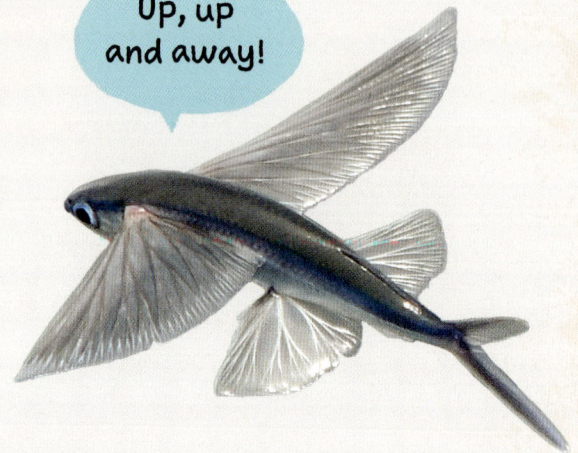

This is no magic trick, even if it does completely fool hungry predators. Flying fish can leap out of the water and glide away through the air, while a kind of deer, the water chevrotain, does the opposite and hides underwater. The effect is the same – the animals disappear from the view of whatever's chasing them.

Where did it go?

Got to get out of here!

The pebble toad has one more trick. This teeny amphibian lives in hilly areas, and when a large, scary spider approaches, **IT TURNS ITSELF INTO A BOUNCY BALL!**

Dinner is served!

Not today!

Better to fall down a mountain than be a spider's dinner!

First, the toad tucks its head, arms and legs against its body. Then, because the toads live on steep slopes, they simply roll downhill out of danger, gathering speed as they go and leaving the spider far behind.

ARMOURED ANIMALS

People have built castles to keep their enemies out for centuries, but animals came up with the idea long before we did.

No one is getting into my castle!

Oh, mussel, your shell is beautiful!

Thank you, urchin, you're not so bad yourself.

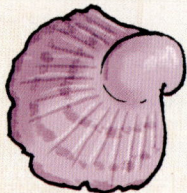

You just try getting past my bony-plated bum!

Whenever you go to the seaside, you'll see plenty of animals with shells, from mussels and urchins to oysters and crabs. And if you go to Australia, you might see wombats, which have **BONY PLATES IN THEIR BUMS.** If threatened, they can stick their heads into their burrows and rely on their reinforced rear to protect them from harm.

But the most impressive of all armoured creatures are tortoises. Their shell is like a fortress that covers their whole body, except for holes allowing their legs and head to poke out. And if a predator makes an unwelcome appearance, the tortoise can pull everything inside its shell, leaving its enemies with nothing but a **TOOTH-SHATTERINGLY TOUGH** suit of armour to contemplate.

How do you get into this thing?

I wouldn't bother trying! You'll break your beak!

Tortoises have some rivals for the title of toughest animal. The ironclad beetle's armour is so robust that it can survive a car driving over it. And the pangolin, a scaly mammal found in Asia and Africa, is so well protected by its scales that it is being used as the inspiration for body armour for people!

They may look like feathers, but they are super-tough scales!

I'm so tough, even cars can't run me over!

PLAYING DEAD

One weird way to avoid being eaten is to be **SO DISGUSTING** that no self-respecting predator would want to deal with you.

This is what the Virginia opossum does. If it finds itself in a tight spot, it rolls onto its side and **PRETENDS TO BE DEAD.** The opossum is a great actor and is really quite convincing – it even slows its breathing right down. Most predators like their food to be fresh, so playing dead puts them off.

Oh no! Here comes a wolf! Got to play dead ...

But just to be double sure, the opossum takes extra measures to deter predators. It drools and its tongue lolls out. Worse still, **A HORRIBLE, STINKY, GREEN FLUID** leaks from its bottom. In other words, it makes itself so repulsive that even the hungriest of predators think twice before tucking into such a nasty meal.

Once its enemy has retreated, the opossum springs to its feet, cleans itself down and gets back to its business.

Ew! I'm not eating this stinky dead thing!

Ha ha! I may be stinky but I'm not dead!

Eurgh, I'm going to be sick.

Scientists have discovered that the reason why animals like the opossum play dead is because lots of predatory animals are attracted to movement. It takes nerves of steel to stay still when something scary is approaching, but if you can't hide or outrun your enemy, freezing might mean it'll pass right by.

25

FIGHTING BACK

Check out my fangs!

Tarantulas – the hefty, hairy spiders found in warmer parts of the world – are pretty **FEARSOME** to look at. They're about the size of your hand and they have sharp, curved fangs that could BITE right through an unwary finger.

I see my next snack!

You might not think that there'd be many animals who'd mess with them. But surprisingly, quite a few creatures, including lizards, birds and foxes, see them as a **TASTY MEAL.**

Any hopeful spider-snacker will need to have its wits about it because the tarantula has a special attack. When its enemies get too close, the tarantula **FIRES HAIRS**, called urticating hairs, out of its bum – straight into their face.

Being hit with bum hair is no joke. Each hair is like a tiny arrow, coated with venom that'll make any would-be attacker think twice about a tarantula takeaway in future.

Agh! Bum hair arrows! That hurts!

Take that!

I may look soft and furry, but these hairs can cause a rash!

TALES FROM THE LAB

In the 1970s, tarantulas were becoming popular as pets. Thought to be safe to humans, they seemed to be causing rashes. So to be sure, in 1972, a spider scientist firmly rubbed the hairy back of a tarantula against his arm and, sure enough, an itchy rash came up. All in the name of science!

Hey birds, will you be my friends?

Lunch is served!

MOBBING

Imagine for a moment that you're a baby buffalo, scampering around in the sunshine on the plains of Africa. You trot past a thicket of trees when suddenly a lion bursts from cover and comes charging towards you with a hungry look in its eye. You're in **TROUBLE!**

Luckily, buffaloes live in big groups. Your mum – and most likely, the whole extended buffalo family – will be keeping watch. The lion might be thinking of you as an easy meal, right up until the moment when half a ton of very angry mama appears at your side.

Not so fast! You go and stand behind me, honey.

You're joined moments later by dozens of your friends and suddenly this has turned into a **VERY BAD DAY** for the lion. Faced with an army of huge, powerful and decidedly furious buffaloes, armed with **WICKED HORNS,** the lion has to run for its life.

You've had a lucky escape, but it just goes to show that there's strength in numbers. When one buffalo is threatened with danger, its friends come to the rescue.

My tummy's grumbling ...

Get out of here!

Go away!

TALES FROM THE LAB

You might not think that an animal as large as a sperm whale would need to worry about predators, but killer whales can sometimes threaten young calves. Marine biologists have seen adult sperm whales gather in a circle to protect their little ones, and thrash with their tails to deter the killer whales.

GLOSSARY

Behaviour the way in which a person or animal acts

Blend in to look or seem the same as the surroundings and not be easily noticeable

Burrow a hole in the ground dug by an animal to live in

Camouflage the way that the colour or shape of an animal or plant appears to mix with its surroundings, in order to prevent it from being seen and attacked

Marine biologist a scientist who studies animal, plant and microscopic life in oceans

Masquerade the act of wearing a disguise and pretending to be something else

Mimicry the act of copying the sounds or behaviour of a particular animal

Mobbing a behaviour where a group of prey approach, scare and sometimes attack a predator

Poisonous uses a harmful or deadly substance in order to defend itself

Predator an animal that hunts other animals for food

Prey an animal that is hunted or caught for food, usually by another animal

Strategy a way of doing something or dealing with something

FURTHER INFORMATION

Books

Science-ology!: Zoology
by Anna Claybourne (Wayland, 2023)

How Not to Get Eaten: More than 75 Incredible Animal Defenses
by Josette Reeves (DK, 2022)

Animals in Disguise by Michael Bright (Wayland, 2023)

Websites and videos

Explore zoology, the study of animals though articles and games:
www.amnh.org/explore/ology/zoology

Watch this video to learn more about the pearlfish
and the sea cucumber:
https://youtu.be/dOoZ6wHiSnI

See how the pebble toad bounces down a mountain to escape a spider:
https://youtu.be/CRaGZnnepSA

Watch this video to see springboks stotting:
https://youtu.be/qr5Sru8gGSk

INDEX